CONTENTS

Words in the glossary appear in **bold** type the first time they are used in the text.

SNEAKY SNAKES

There are more than 3,000 species, or kinds, of snakes in the world. Some are very small, and others are huge. Many have different tricks to help them survive.

Snakes have special ways of finding **prey**. They use their tongue to smell the air around them. They also have openings called pit organs in front of their eyes. Pit organs sense the heat a warm body gives off. As they move along the ground, the bones in their lower **jaw** feel the ground vibrate, or shake, as their prey moves.

FUN FACT

There are no snakes in Antarctica, Ireland, Greenland, or New Zealand!

Snakes have nostrils, or nose holes, just like humans do. However, they don't use their nostrils to smell. They only use them to breathe.

THE PUFF ADDER

Puff adders can be found in parts of Africa and the Arabian **Peninsula**. They eat small animals such as mice, lizards, frogs, and birds. They have a special trick to help them catch frogs.

Normally, a snake flicks its tongue quickly in and out of its mouth to smell the air. When a puff adder sees a frog, it flicks its tongue much more slowly. To the frog, the tongue looks like a worm it wants to eat. It gets close to the snake—and then it's the one that gets eaten!

FUN FACT

When animals use their tongue to trick prey into getting close enough to catch, it's called lingual luring.

Puff adders are mainly found in southwestern Saudi Arabia and Yemen, as well as most of Africa.

THE MOCK VIPER

A viper is a kind of **venomous** snake. It uses this venom to kill its prey and to warn away predators. The mock viper, however, is not venomous. This makes it **vulnerable** to large predators. To trick predators into leaving it alone, the mock viper makes itself look like a true viper.

The head of a mock viper is the same shape as the head of a true viper. Sometimes this is enough to make predators steer clear. If one gets too close, the mock viper can change the shape of its **pupils** to look more like those of a true viper.

Mock vipers (top) have round pupils. True vipers (bottom) have **slit** pupils. When they need to fool predators, mock vipers can also make their pupils look like slits.

THE CUBAN BOA

A boa constrictor, also simply called a boa, is a kind of snake that uses constriction, or squeezing, to kill its prey. The Cuban boa is the only species of boa that lives on the island country of Cuba.

Cuban boas, like most snakes, often hunt alone. However, they have figured out that sometimes, working together is a trick that helps them all get more food. They're one of the only species of snake in the world that uses teamwork to hunt.

The Cuban boa is the largest native predator in Cuba.

TEAMING UP

Cuban boas prey on fruit bats, which live in caves in Cuba. At night, the bats fly out of their caves to hunt. The boas move into the cave while the bats are gone. They spread out in a line and hang from the ceiling.

When the bats come back home in the morning, they run into the line of boas. If there were only one snake, the bats could easily avoid going near it. However, a line of snakes is harder to avoid. Hunting like this helps all the boas catch at least one bat.

Cuban boas hunt bats in caves such as this one.

FUN FACT

Cuban boas are one of very few snake species scientists have been able to watch as they hunt in the wild. There might be more snakes that hunt together, but scientists haven't seen them do it yet.

THE YELLOW-BELLIED SEA SNAKE

About 70 species of snakes, including the yellow-bellied sea snake, live in the Indian and Pacific Oceans. Their bodies have adapted, or changed over time, to help them survive better in the water.

All snakes shed, or lose, their skin between 4 and 12 times per year. On land, snakes help the skin come off by rubbing up against a rock, tree, or other hard object. In the water, where there are fewer hard objects, yellow-bellied sea snakes use a trick: They tie themselves in a knot and use their own body to rub off their skin!

FUN FACT

A snake sheds its skin because its body grows, but its skin doesn't. A new layer of skin grows under the old layer, and the snake sheds the old, outgrown layer.

The yellow-bellied sea snake also ties itself in a knot to clean its skin.

THE GOLDEN TREE SNAKE

While some snakes swim, others fly! Scientists have found five different species of flying snakes. The largest is called the golden tree snake.

The golden tree snake flies by stretching out its ribs. This makes the snake's body flat and wide, so it catches the air like a kite. It controls its flight by moving its body through the air in the same way a snake moves its body on the ground. Scientists aren't sure exactly why these snakes fly, but it might be to help them get away from predators quickly.

The golden tree snake is **endemic** to South and Southeast Asia.

FUN FACT

Even though they're called "flying" snakes, these species don't really fly. Instead, they **glide**. When they glide, they can move at up to 25 miles (40 km) per hour.

THE HOGNOSE SNAKE

The hognose snake gets its name because the tip of its nose looks a little like a pig's nose. This nose isn't just for show! The snake can use it to dig itself into the sand.

Hognose snakes mostly eat toads. The toads can stop other snakes from eating them by swallowing air, so they puff up and become too large for the snake to eat. However, hognose snakes have a trick for this. They use their teeth to pop the toad!

The hognose snake can
use its nose like a shovel.

FULL OF TRICKS

The hognose snake has other tricks, too. When a predator comes near—including a human—it will flatten out its head and neck to look like a cobra's hood. It will also hiss like a cobra does. If that doesn't scare away the predator, the snake will attack, but it won't always bite. Often, it will keep its mouth closed.

If all else fails, the hognose snake will play dead. Since its predators don't want to eat a dead snake, they'll usually leave it alone. The snake is so good at this trick that it won't even move if it's picked up!

A cobra (bottom) is a very deadly kind of snake. The hognose snake (top) makes itself look a little like a cobra to trick predators into thinking it's also deadly.

FUN FACT

Sometimes humans kill hognose snakes because the snake's tricks are so good. People think it really is deadly. Hognose snakes are harmless to humans though!

GLOSSARY

endemic: Found only in a certain place.

glide: To descend gradually without enough power for flight.

jaw: One of the two bones of the face where teeth grow.

peninsula: A narrow piece of land that extends into water from the mainland.

prey: An animal that is hunted by other animals for food.

pupil: The black part in the center of an eye that lets in light.

slit: A long, narrow opening.

venomous: Able to produce a liquid called venom that is harmful to other animals.

vulnerable: In an unsafe position.

FOR MORE INFORMATION

Books

Hand, Carol. *Snakes*. Minneapolis, MN: Abdo Reference, 2023.

Lundgren, Julie K. *Snakes*. New York, NY: Crabtree Seedlings, 2022.

McDonald, Amy. *Snakes*. Minneapolis, MN: Bellwether Media, 2021.

Websites

DK Find Out!: Snakes
www.dkfindout.com/us/animals-and-nature/reptiles/snakes
Learn more about different kinds of snakes.

Kiddle: Snake Facts for Kids
kids.kiddle.co/Snake
This website includes some cool photos of snakes.

US Forest Service: Snakes
www.fs.usda.gov/visit/know-before-you-go/snakes
Read more about what to do if you see a snake, how to avoid being bitten, and more.

INDEX